YOUR KNOWLEDGE HAS VALUE

Bibliographic information published by the German National Library:

The German National Library lists this publication in the National Bibliography; detailed bibliographic data are available on the Internet at http://dnb.dnb.de .

Imprint:

Copyright © 2017 GRIN Verlag, Open Publishing GmbH
Print and binding: Books on Demand GmbH, Norderstedt Germany
ISBN: 9783668411524

This book at GRIN:

http://www.grin.com/en/e-book/354931/physico-chemical-analysis-of-pycnanthus-pycnanthus-angolensis-seed-oil

Adekunle Jelili Olaoye

Physico-Chemical Analysis of Pycnanthus (Pycnanthus Angolensis) Seed Oil

GRIN Publishing

GRIN - Your knowledge has value

Since its foundation in 1998, GRIN has specialized in publishing academic texts by students, college teachers and other academics as e-book and printed book. The website www.grin.com is an ideal platform for presenting term papers, final papers, scientific essays, dissertations and specialist books.

Visit us on the internet:

http://www.grin.com/

http://www.facebook.com/grincom

http://www.twitter.com/grin_com

PHYSICO-CHEMICAL ANALYSIS OF PYCNATHUS (*Pycnanthus angolesis*) SEED OIL

OLAOYE, A.J

Chemistry Department Ladoke Akintolan University Of Technology, Ogbomoso, Oyo State, Nigeria.

ABSTRACT

This study was based on physico-chemical analysis of the oil content of Pycnanthus seed for its potential and industrial uses. Seeds were collected from Pycnanthus tree at Awba-dam, UI Ibadan. The results obtained are oil extract 48%, the specific gravity 0.978, refractive index 1.4521, melting point 50^0C, the saponification value 245.44%, acid value 6.21mg(OH)/g oil, peroxide value 16.12mEq/kg and ester value 239.23. The iodine value of seed oil which placed the oil in the non-drying group was 84.94. Fatty acid composition of the seed oil showed the oil to be rich in lauric acid 64.72%, palmitic acid 13.97%, capric acid 4.99%, capryclic acid 4.40%, myristic acid 4.33% and stearic acid 1.93%. It also contained oleic acid 3.26% and linoleic acid 2.40%. The other fatty acids present in the oil are palmitoleic acid, Linolenic acid and Llignoceric acid. The seeds of the plant could be a source of industrial raw material.

Key words : Pycnanthus angolesis, Specific gravity, Iodine value, Refractive index and Fat content

Table of Contents

INTRODUCTION – PYCNANTHUS ANGOLENSIS

"Akomu" tree is biologically called *Pycnanthus Angolenis*. It is called African nutmeg in English, Akwa-mili in Igbo, Lunaba in Luganda, Calabo in Spanish. The trade name given to this tree is Ilomba while it is called Akomu in Yoruba. African nutmeg belongs to the family of Myristicacease (Center Technique forestier Tropical 1961). Parts of the plant are widely reported to have several medicinal value (Agyare et al., 2009). The leaf and bark are used to treat toothache and the sap of of the plant is applied topically to arrest bleeding (Abbiw, 1990).

Pycnnthus angolensis is a tree of about 25 – 35 (40) m high and 60 – 100 (150) cm in diameter, occasionally more, evergreen, bole straight, cylindrical, without buttresses, bark grey, longitudinally fissured, flacking in patches , in old trees, slash reddish, exuding a sticky, honey – coloured sap turning red. The leaves are so often eaten by insects that this is a characteristic feature (Abbiw, 1990).

Fruits (often occuring within the flowers) ellipsoid or almost spherical drupe shape, 2.5 – 3.8cm long and 1.9 – 3.2cm in diameter, often indense clusters at the base of the twigs, opening by 2 values and exposing a solitary black seed with a bright red arill much branched at the apex. The generic name, Pycnanthus is derived from Greek, the literary meaning is 'dense flowers' and it refers to the numerous flowers crowded together. It is distributed across the native such as Benin, Congo, Cameroon, Cote d'Ivore, Nigeria, Senegal , Uganda (Abbiw, 1990)

The evergreen tree is the monoecious, with the asexual flowers on different parts of the same branch. In its natural habitat, the flowers are produced in October and November, at the same time on the tree until about February. Deliscence takes place on the tree, but many of the fruit clusters fall unopened (Dupuy, 1993).

FUNCTIONAL USES

Pycnanthus angolensis tree serves a lot of functions. It is used as fuel where the seeds burn like candles and seed oil is an illuminant in West Africa(Arbonncer, 2004). As lipids, the oil is extracted from the seeds and used in making soap. Medically, bark decoction is an emetopurgative and can act as an antidote to poisonig; used in treating leprosy and, if pounded, used to treat stomachache (Keay et al., 1964). Its Sap acts as a septic (arrests bleeding). Leaf and bark help to relieve tootache (Abbiw, 1990). Seed fat and probably leaf juice is used in treating thrush. Root infusion acts as an anthelmintic (Abbiw, 1990). It is used as timber whereby the wood obtained is grayish white or tinged with pink. The wood is light, very soft, of medium nervosity and shrinkage. Its natural durability is low, but it is easily machined. During seasoning the wood sometimes warps (Udeozo et al., 2011).

This easily worked, and straight-gained wood is used for veneer peeling, panels, furniture frames, box making and minor joinery. In Cameroon, It is split into rough planks for house building and roofing materials (Arbonncer, 2004)

MATERIALS AND METHODOLOGY

SAMPLE COLLECTION

Pycnanthus seeds were obtained from the Pycnanthus tree at Awba-Dam, University of Ibadan, Oyo State, Nigeria in the Month of June, 2015.

SAMPLE PREPARATION

The seeds were washed with water to remove the unwanted materials, afterwards, the seed were cut into pieces so as to aid effective drying. They were then sundried in the sunlight for two weeks and again dried oven at 40^0C for an hour. The seeds were grounded into powder and kept in a sealed container.

DETERMINATION OF PHYSICAL AND CHEMICAL PROPERTIES OF PYCNANTHUS SEED OIL

The parameters analysed were melting point specific gravity, saponification value, acid value, iodine value, refractive index, lipid.

CRUDE FAT DETERMINATION

SOLVENT SELECTION

The idea solvent for lipid extraction is one, which could completely extract all the lipid from the sample while leaving all other component behind (Rosenbery, 1998).

From this research work, hexane is chosen because it is non-polar, non-toxic and easily be removed by evaporation (Ibitoye, 1996).

EXTRACTION METHOD

The batch solvent extraction (cold extraction) was used for this work.

BATCH SOLVENT EXTRACTION

40 g of the dried sample was weighed into a separatory funnel and about $250cm^3$ of the solvent (hexame) was added. The separatory funnel containing the mixture was vigorously shaken intermittently.

It was then allowed to stay for $24 - 48$ hours to ensure adequate extraction and settlement.

The acqueous phase was then decanted off and the solvent was allows to ecaporate leaving behind the lipid.

% Fat= $\dfrac{\text{loss in weight}}{\text{Weight of fresh sample}}$ X 100

$$= \dfrac{W_1 - W_2}{W_1} \text{ X } 100$$

Where:

W_1 =weight of fresh sample=40g

W_2 =weight of sample after=20.8g

The mass of the lipid was measured to determine the amount of the lipid extracted.

(Morris, 1999)

MELTING POINT DETERMINATION

The fat sample was melted and drawn into a capillary tube. The tube containing, the sample was then placed in a refrigerator between 5 °C - 10 °C for 18 hrs for the oil to solidify. The sample was then placed in amelting point apparatus and observed until it began to melt. The temperature range at which melting occurred was noted and recorded as the melting (AOCS, 1994).

SAPONIFICATION VALUE

2g of oil was weighed into a conical flask, 25cm³ of 0.5M alcoholic KOH was added. A blank was prepared by putting 25cm³ of the alcoholic KOH in a similar flask reflux condenser was fitted to the flask containing the mixture, this was heated in a water bath for one hour, swirling the flask from time to time. The flask was the allowed to cool a little and the condenser was washed down with a little distilled water, the excess KOH was titrated with 0.5M HCL using phenolphthalein indicator.

The saponification value was calculated from the difference between the blank and the sample titration.

Saponification value = $\dfrac{(a\text{-}b) \times f \times 28.05 \times 100}{\text{Weight of sample}}$

Where: a = titre value of sample

b = titre value of blank

f = factor of 0.5M HCL

28.05 = mg of KOH equivalent to 1cm³ of 0.5M HCL

Weight of sample = 2g (Morris, 1999)

ACID VALUE

0.1g of the oil was dissolved in 2.5cm³ of 1: 1v/v ethanol: diethyl ether solvent and titrated with 0.1N sodium hydroxide while swirling using phenolphthalein as indicator.

The acid value was calculated using the formula:

Acid value = $\frac{5.61 \times N \times V}{W}$

Where: N = the normality of sodium hydroxide

V = the volume of sodium hydroxide in cm^3

W = weight of the sample = 0.1g (Morris, 1999)

ESTER VALUE

This was obtained by finding the difference between the saponification value (S.V) and Acid value (A.V) (Morris, 1999)

PEROXIDE VALUE

5g of the oil sample was dissolved in $30cm^3$ of a slovent mixture consisting of 60% glacial acetic acid and 40% of chloroform. $0.5cm^3$ of a saturated solution of potassium iodide was added. The flask was shaken until the solution became clear. After 2 minutes from the time of addition of KI, $30cm^3$ of distilled water was added and titrated with 0.01N sodium thiosulphate solutions. It was then shaken vigorously to remove the last traces of iodine from the chloroform layer.

Peroxide value (milliequivalent/1000g) = $\frac{ML \times N \times 1000}{W}$

(millimole/1000g) = $\frac{0.5 \times ML \times N \times 1000}{W}$

Where: ML = titre value

N = normality of $Na_2S_2O_3$ solution

W = weight of the oil sample = 5g (Morris, 1999)

IODINE VALUE

1g of oil was weighed into a conical flask. $10cm^3$ carbon tetrachloride was added, then $20cm^3$ of wijs solution was added and the flask was covered, mixed and allowed to stand in the dark for thirty minutes. 10% potassium iodine was prepared by wighing 10g and dissolved it in 100cm^3 of distilled water. Then 15cm^3 of the prepared 10% KI solution and 100cm^3 distilled water were added to the content in the flask. It was mixed thoroughly and titrated against 1.0N thiosulphate solution. Starch indicator was uded and blank determination was carried out under the same condition.

IODINE VALUE (I.V) = $\underline{12.69 \times N \times (V_2 - V_1)}$

W

Where: Weight of oil sample (W) = 1g

V_1 = Volume of thiosulphate used in blank

V_2 = Volume of thiosulphate used in test

N = Normality of thiosulphate solution (Morris, 1999)

DETERMINATION OF SPECIFIC GRAVITY OF THE SOIL

Specific gravity of oil was determined at room temperature. 10cm^3 of the oil sample was weighed as W_2. 10cm^3 of distilled water was also weighed and the weight was recorded as W_1

Specific gravity = $\dfrac{W_2 + 0.00064 \, t^1}{W_1}$

$W_2 = 0.4266$

$W_1 = 0.4448$

t = temperature at which oil is weighed (45^0)

$t^1 = t - 15.5^0c$ (Morris, 1999)

REFRACTIVE INDEX

Refractive index of the oil was done using abbe refractometer the surface of the prisms was cleaned up with either. 2 drops of the oil was applied at the lower prism and the prisms were closed up. Water was passed through the jackets at 45^0c.

The jacket was adjusted for reading to be taken. (Morris, 1999)

DETERMINATION OF FATTY ACID COMPOSITION

The oil sample was air dried by blowing the air gently on to it. 1.0g of the oil was weighed into the beaker and heated in a borosilicate beaker container at 140^0F with pump running to allow homogeneity of the sample. Some drops of the acid were added to the oil in the container in a fairly fast manner for the proper distribution. The sample was homogenized while the acid was being added with the aid of the mixer.

About $3cm^3$ of methanol was measured into a precleaned beaker. The heater and the pump were off to allow the methanol to be added in a fast way. The mixture was mixed properly and the fan was allowed to blow the fumes away. The mixture was covered properly and temperature was allowed to drop to ambient temperature (65^0F), and the pump and fan were on occasionally for about four hours. After cooling, the sweet fragrance ester was decanted into a clean borosilicate container before injecting into the Gas chromatography. (igwe et al, 2005)

RESULT AND DISCUSSION

RESULT

Table 1: Result of physical properties of Pycnanthus seed oil

PROPERTIES	COMPOSITION
Specific gravity at 25^0C	0.978
Refractive index at 45^0C	1.4521
Melting point (^0C)	50^0C
Fat content (%)	48

As shown in table 1, specific gravity of Pycnanthus seed oil at 25^0C was 0.978, being lower than 0.9962 at 25^0C obtained by (Artherton and meara, 1939) for Karela seed oil. Also, this value is higher than the specific gravities obtained for Baobab seed 0.8755 (Ojo, 2006) , 0.922 Kombo kernel (Nagre *et al.*, 2011), 0.36 Pycnanthus angolesis (Chigozie *et al.*, 2014 and Udeozo et al.,2015) . It indicates that the oil is less than water.

The refractive index of the oil found in the investigation was 1.4521 at 40^0c, being lower than that obtained by (Artherton and meara, 1939) for Karela see oil, which was 1.4985 at 25^0c, 1.461 for Kombol kernel (Nagre *et al.*,2011)

The melting point of the oil was found to be 50^0c, being close to the value 51^0c reported elsewhere (http:// database. Prota. Org/) and higher than 43 that obtained by (Nagre et al., 2011) for Kombo kernel.

The solvent extract of Pycnanthus seed yield 48% oil, which falls within the range 45 – 70% reported on (http:// database Prota. Org/). This value also is higher than the value 26% reported by (Ajayi, 2004) for the same oil.

Table 2: Chemical characteristics of pyscnanthus seed oil

Characteristics	Composition
Saponification (%)	245.44
Acid value (mg(oH)/g oil)	6.21
Ester Value	239.23
Peroxide value	16.12
Iodine value	84.94

As given in Table 2, the saponification value of Pycnanthus seed oil was 245.44, being lower than the value 252.11 for the same oil reported by (Ajayi, 2004), but higher than 238.0 reported by (Nagre *et al.*,2011). This value, when comapred with the values obtained for some vegetable oils ranging from 188-196, is higher. However, there are some vegetables with higher saponification values such as coconut oil 253, palm kernel oil 247 and butter fat 225. It has been reported by (Pearson,1981) that oils with higher saponification values contain higher proportion of lower fatty acids. Therefore the value obtained for Pycnanthus seed oil indicated that the oil contained higher proportion of lower fatty acids (Aremu *et al* 2006).

Also the saponification value obtained is significantly different from those of edible oils ranging from 188-198 set by the international codex standard for edible oil, (Pearson, 1981 and NIS, 1992). These higher values also buttress the fact that it is suitable for making soap. High saponification values of fats and oils are due to predominantly high proportion of shorter car bon chain length of fatty acids (Kirk and Sawyer,1991).

The acid value of the pychnanthus seed oil was estimated to be 6.21, and was thus lower than the value 14.31 and for the same oil reported by (Ajayi, 2004 and Nagre *et al.,*2011). The acid value of the oil is significantly different from those obtained for some vegetables oil 0.4 – 1.2 (Otunola et al. 2009), this value when compared to the standard range of 0.5 – 1.5 (Pearson, 1981) shows that the oil will not go rancid easily.

The ester value of the oil was calculated to be 239.23 from the acid and saponification values. Ester value of this experiment was higher than the value 213.69 repoted by (Nagre *et al.,*2011) for the same oil.

The peroxide value of Pycnanthus seed oil was found to be 16.12. This value is higher than 15.90 and 9.9 reported by (Ajayi, 2004 and Nagre *et al.,*2011) for the same oil. It is also higher than those obtained for the varieties of Karela seed 6.13 – 8.50 (Ali *et al.* 2008) peroxide value is an indication of the product of primary oxidation. It measures the degree of oxidation. The higher the peroxide the value of oil, the more undesirable the oil will be in flavour or taste.

The iodine value of Pycnanthus seed oil was estimated to be 84.94, and thus lower than the value 85.00 and 70.6 reported by (Ajayi *et al,* 2004 and Nagre *et al.,*2011) for the same oil.

The iodine value obtained in this study is lower than the values 91 – 119.4 reported for different vegetable oils (Otunola *et al.* 2009). This value indicates that Pycnanthus seed oil has a lower degree of unsaturation and that it is not placed among the oils in the drying group (Ajayi *et al.* 2004)

Table 3: Fatty acid composition of Pychnanthus seed oil.

Fatty acid	Nomenclature	Type	Relative composition (%)
Caprylic acid	C8:0	Saturated	4.3990
Capric acid	C10:0	Saturated	4.9937
Lauric acid	C12:0	Saturated	64.7175
Myristic acid	C14:0	Saturated	4.3277
Palmistic acid	C16:0	Saturated	13.9690
Palmitoleic acid	C16:1	Unsaturated	0.0010
Stearic acid	C18:0	Saturated	1.9325
Oleic acid	C18:1	Unsaturated	3.2556
Linoleic acid	C18:2	Unsaturated	2.3989
Linolenic acid	C18:3	Unsaturated	0.0005
Arachidic acid	C20:0	Saturated	ND
Behenic acid	C22:0	Saturated	0.0017
Erucic acid	C22:1	Unsaturated	0.0001
Lignoceric	C20:0	Saturated	0.0027

Fatty acid compositions of Pycnanthus seed oil was determined by GLC

(Gas Liquid Chromatography), and were presented in table 3. The result shows that both saturated and unsaturated fatty acid were found in the oil. GLC data revealed that pycnanthus seed oil contained a higher amount of saturated fatty acid (94%), while unsaturated fatty acid was found to be 6%. The fatty acids profiles evaluated in this study are not in complete agreement with the works reported (Ajayi *et al.* 2004) for the same oil. The fatty acid composition alters with the variety, oil and climatic conditions (White and dietenberger, 2001).

However, Pycnanthus seed oil displayed a higher degree of saturation as compared to Karela seed oil, with a higher degree of unsaturated fatty acid (Ali *et al.* 2008). Also, GLC data revealed that the degree of unsaturated is lower than that of various edible oils reported by (Otunola *et al.* 2009). Therefore, Pycnanthus seed oil cannot be placed among the edible vegetable oil since linoleic and oleic are major constituents of most edible vegetable oils (Compaore *et al.* 2011), the presence of linoleic and oleic acid in a very small amount was responsible for the low iodine value, and is also indication that the shelf life of the oil might be long. Vegetable oils that contain fatty acids with conjugated double bonds, such as tung oil, are valuable drying agents in paints, varnishes and inks (Agyare, 2009). Pycnanthus seed oil reported here in was enriched in lauric acid (64% of the total fatty acid) with no conjugated double bonds and not used commercially in paints and ink but suitable for making soap and candle.

REFERENCES

Abbiw, D. (1990). Useful plants of Ghana. *Intermediate Technology Publication and the Royal Botanical Gardens*, Kew.

Ajayi I.A., Adebowale K.O., Dawodu F.O and Oderinde R.A. (2004). A study of the oil content of Nigerian grown *Monodora myristica* seeds for its nutritional and industrial applications. *Pakistan Journal of Scientific and Industrial Research*, 47:60 – 65.

Ali M.A., Sayeed M.A., Reza M.S., Yeasmin Mst.S. and Khan A.M. (2008). Characteristics of seed oils and nutritional compositions of seed from different varieties of *Momordica Charantia Linn*. Cultivated in Bangladesh, *Czech Journal of Food Science*, 26:275 – 283.

Aremu M.O., Olanisakin D.A., Bako and Madu P.C. (2006) Compositional studies and Physico-chemical Characteristics of cashew nut flour. *Pakistan Journal of Nutrition*, 5:328 – 333.

Center Technique Forestier Tropical (1961). Monographie de I' II omba, Pycnanthus angolensis (Welw) Warb. Publication No. 20 du center Technique Forestier Tropical. Norgent – sur – Marne. (Seine).

Depuy B, and Mille G. (1993). Timber Plantations in the humid tropics of Africa. *FAO Forestry paper.* 98.

Http://database. Prota. Org/PROTAhtml/ Pycnanthus %20 angolensis_ En. Htm. (2001).

Nagre, R.D., Oduro, I., and Ellis, W.O (2011). Comparative physico- chemical evaluation of kombol kernel fat produced by three different processes. *African Journal of Food Science and Technology* 2(4): 083 – 091

Chigozie,M.E., Chukwuma,S.E. and Augustine,N.E (2004). Determination of physical and phytochemical constituents of some tropical timber indigenous to Niger Delta area of Nigeria.*European Scientific Journal* 10(18): 247 – 270

Udeozo, I.P., Ejikeme, C.M., Eboatu, A.N. and Arinze, R.U (2015). The Efficacy of Pycnanthus angolesis Timber. An essay of its properties, Chemical Constituents and Functional Group Analysis. *Global Journal of Biotechnology and Biochemistry*, 10(3): 121 – 125

Ibitoye O.A(1996) Laboratory Manual on basic qualitative chemical methods and calculations.Analytical Laboraty Department, Federal University of Technology, Akure

Igwe C.C., Yayi E. and Moudachiou M.(2005).Chemical Constitutes of the solvent Extracts and Hydrodistilled Essential oils of African Nutmeg (Monodora Myristica) and Tumeric (Curcuma domestica) from south west Nigeria.*Nigeria Food Science Journal*, 23:21 -33.

Ihekoronye A.I. and Nogddy P.O. (1991). Integrated Food Science and Technology for the tropics: 1st ed. Oxford; Basil Blackwell Ltd.

Morris, B.J. (1999). The chemical Analysis of food and food products: 3rd ed. India: CBS Publisher and Distributors.

Nigerian Industrial Standard (NIIS), (1992). Standard for Edible Vegetable Oil.12.

Ojo, T.O. (2006). Proximate and Physico chemical analysis of Kapok seed oil.Applied Chemisty Department, Federal Polytechnic, Ede.

Otunola G.A.,Adebayo G.B.and Olufemi O.G. (2009). Evaluation of some Physico-Chemical parameters of selected brands of vegetable oils sold in Ilorin metropolis. *International journal of Physical Sciences*, 4;327-329.

Pearson, D. (1981).The chemical analysis of foods: 7th edition, churchhill livingstone. 44-496.

Rosenbery, F. (1998). The book of edible nuts. Walker and Co New York.S

Arbonncer, M. (2004). Trees Shrub and Lianas of West African Dry Zones. Grad, Magrat Publisher, 574.

Udeozo, I.P., Eboatu, A.N., Arinze, R.U and Okoye, H.N (2011). Some fire characteristics of fifty-two Nigerian Timber. *Anachem Journal*, 5(1): 920-927.

Agyare,C., Asase, A., Lechtenberg, M., Niehues, M., Deters, A and Hensen (2009). An Ethnopharmacological use of medicinal plants used for wound healing in Bosomtwi-Atwima-Kwanwoma area Ghana. *A Journal of Ethnophamacology,* 125:393-403

Atherton, D and Meara, M.L (1939). Fatty acids and Glycerides of Some Myristica Fats. *Journal of Society of Chemical Industry London*, 353-357.

Compaore, W.R., Nikiema, P.A., Bassole, H.I.N., Savadogo, A., Movecovcou, J., Hounhoug, D.J and Traore, S.A (2011). Chemical composition and Anti oxidative Properties of Seeds of Moringa oleifera, pupls of parkia biglobosa and Adansonia digitata Commonly Used in Food Fortification in Burkinafaso. *Current Research Journal of Biological,* 3(1):64-72.

White, R.A and Dietenberger, M.A (2001). Wood production: Thermal, Degradation and Fire. Encyclopedia of Materials Science and Technology. E/ Science Limited, *Washington, D.C.,* 9712- 9716.

Keay, R.W.J., Onochie, C.F.A and Stanfield, D.P (1964). Nigeria Trees, *Department of Forest Research Publishers Ibadan,* 1:38- 265.

Kirk, R.S and Sawyer, R (1991). Pearson's Composition and Analysis of Food, 9th edition, *Addison Wesley Longman Limited,* England, 9-29, 608 – 640.

YOUR KNOWLEDGE HAS VALUE